KTM EQUINE LLC

A Structured Durability Architecture for the Two-Year-Old Thoroughbred

The KTM Model Program

Kimberly Schaffer-Marrs
KTM Equine LLC

First Edition · 2026

A Structured Durability Architecture for the Two-Year-Old Thoroughbred

KTM Model Program

Kimberly Schaffer-Marrs
KTM Equine LLC

First Edition — 2026

Printed in the United States of America

Subject: Thoroughbred racehorse training · Equine sports medicine · Skeletal adaptation

FOREWORD

The modern Thoroughbred racing industry stands at an inflection point. Performance expectations have never been higher, yet durability concerns have never been more scrutinized. The two-year-old season remains both the greatest opportunity and the greatest structural risk in a racehorse's career.

The KTM Model Program was developed from a single guiding principle: performance and durability are not opposing forces. When sequencing is structured correctly, they reinforce one another.

This book is not an argument against tradition. It is a refinement of it. The twelve-second furlong standard remains intact. Competitive readiness remains central. What changes is architectural discipline.

The pages that follow present a biologically grounded, culturally aligned framework for programming the two-year-old season in a way that protects structural margin, preserves long-term earning potential, and strengthens the three- and four-year-old campaigns that follow.

A Template, Not a Rigid Formula

It is essential to understand what this program is and what it is not. The KTM Model is a structured template — a biologically grounded framework that establishes the sequence, timing, and principles that protect structural margin across the two-year-old season. It is not a rigid formula that ignores the horse in front of you.

Every horse is an individual. Conformation influences how load is distributed through the limb. Temperament influences how a horse handles training stress. Some horses adapt faster than others at the same distance and the same pace. A horse that shows exceptional tolerance at 3/8 may be ready to advance sooner than the template suggests. A horse that shows any sign of heat, filling, or asymmetry after a work needs more time regardless of what week the calendar says it is.

The program gives the trainer a sound biological sequence to work within. The trainer gives the program a set of eyes, hands, and judgment that no calendar can replace. These two things work together — the architecture provides the structure, the horseman reads the horse.

THE KTM PROGRAM — WHAT IT IS AND WHAT IT IS NOT

IS: A biologically grounded template that sequences load, recovery, and competition to protect structural margin across the two-year-old season. IS: A framework that works with veterinary oversight — the horse is in visible daily work throughout every phase, including the Active Resets. IS: A financially sound investment strategy — the cost of two six-week reset blocks is a fraction of the cost of a single preventable injury. IS NOT: A rigid formula. Every horse is an individual. Conformation, temperament, and individual response guide decisions within the framework. IS NOT: A guarantee against injury. It is a systematic reduction of the preventable fatigue-related injuries that account for the majority of racing's structural casualties. The template gives you the architecture. The horse gives you the adjustments. Together, they give you the best possible chance at a durable, competitive athlete.

Kimberly Schaffer-Marrs
KTM Equine LLC · 2026

TABLE OF CONTENTS

CHAPTER ONE

The Structural Opportunity of the Two-Year-Old Season

> ***The two-year-old season is not a risk to be minimized.***
>
> ***It is an opportunity to be engineered.***

The Most Responsive Window in a Horse's Career

The two-year-old season represents the most biologically responsive period in the career of the Thoroughbred racehorse. It is the year in which skeletal modeling is most dynamic, cortical geometry can be influenced most meaningfully, and structural margin can be deliberately expanded through properly sequenced load exposure. This window does not reopen at three. It does not reopen at four. What is built — or left unbuilt — at two shapes the structural reality of every season that follows.

Bone That Responds to Load

At two years of age, bone is highly responsive to mechanical strain. Modeling predominates over remodeling. When high-speed strain thresholds are exceeded appropriately, cortical thickening occurs, cross-sectional area increases, and structural strength improves. This is not theoretical — it is well documented in equine orthopedic research. Young Thoroughbreds subjected to progressive high-speed exercise develop measurably greater cortical thickness in the cannon bone than unexercised age-matched controls. The skeleton, at this age, is actively seeking the signal to build.

However, the same cyclic loading that stimulates adaptation also generates microdamage. Each stride during high-speed galloping produces bending, compression, and torsional forces through the distal limb. Thousands of cycles accumulate in a single breeze. Microdamage formation is not abnormal — it is part of the adaptive process. What determines outcome is balance.

The Balance That Determines Everything

When strain exposure is progressive and followed by sufficient consolidation, modeling and remodeling remain synchronized. Structural margin increases. When escalation is compressed and density is stacked, fatigue accumulates faster than remodeling can resolve it. Structural margin narrows.

THE STRUCTURAL LEDGER

Every training decision in the two-year-old year makes an entry in a cumulative structural account.

DEPOSITS: Progressive high-speed exposure at appropriate intervals. Six-week rest blocks that allow suppressed remodeling to complete. Repetition before escalation.

WITHDRAWALS: Compressed escalation. Stacked density without recovery. Race starts without adequate consolidation time between them.

When deposits consistently exceed withdrawals, structural margin grows. When withdrawals dominate, fatigue debt accumulates — and that debt does not clear when the horse turns three. It compounds.

Speed Is Necessary — Density Is the Variable

High-speed exposure is necessary. Jogging alone does not stimulate sufficient strain magnitude to maximize cortical adaptation. The issue is never speed itself. The issue is escalation velocity and cumulative density. The horse that arrives at the starting gate with a well-adapted skeleton is not the horse that was protected from fast work. It is the horse whose fast work was sequenced correctly — introduced progressively, escalated deliberately, and punctuated by recovery windows that allowed adaptation to consolidate before the next demand was placed.

The Blueprint Phase

The KTM Model Program is built on one central recognition: the two-year-old year is structural programming for the next three seasons. Every furlong, every week, every escalation decision contributes to a cumulative structural ledger. The two-year-old season is the blueprint phase. Architecture must be intentional.

THE KTM CENTRAL RECOGNITION

Performance and durability are not opposing forces.

When sequencing is structured correctly, they reinforce one another.

The two-year-old year is structural programming for the next three seasons.

Durability is not luck. It is architecture.

Research Foundation — Chapter 1

Cortical Adaptation to High-Speed Exercise

Firth, E.C. (2006). Journal of Anatomy, 208(4), 513–526. Firth et al. (2005). NZ Vet J 53(6):377. High-speed exercise during the 2YO year produces measurably greater cortical thickness and cross-sectional area than low-intensity conditioning alone.

Modeling Window and Skeletal Responsiveness

Burns, T.A. (2021). Animals, 11(2), 463. PMC7916178. Rogers, C.W. et al. (2020). Animal Production Science 60:2069. Bone modeling responsiveness peaks in the 2YO year; the geometric advantage of early cortical loading cannot be recovered at a later age.

Cumulative Load Density and Injury Risk

Crawford, K.L. et al. (2021). Animals, 11(1), 270. PMC7910838. Cumulative high-speed exercise volume in the preceding 4 weeks was the primary modifiable risk factor for MSI.

Six-Week Rest Blocks and Structural Consolidation

Whitton, R.C. et al. (2014). The Veterinary Journal, 202(3). PubMed 25296852. Continuous high-speed training suppresses the remodeling cycle. Rest periods allow suppressed remodeling to activate and complete.

CHAPTER TWO

Mechanical Strain and Skeletal Adaptation

Mechanical strain is the language bone speaks. Every structural adaptation is a direct response to load.

Strain Threshold and the Adaptive Response

Bone does not strengthen in anticipation of exercise. It strengthens because of it. Without sufficient strain magnitude, adaptation is incomplete. Osteocytes — the sensor cells embedded within bone tissue — detect deformation under load and send chemical signals recruiting either bone-building osteoblasts or bone-removing osteoclasts. This mechanism only activates above a minimum threshold. In vivo measurements of the cannon bone at racing speed show peak compressive strains of 4,000 to 5,000 microstrain — far above what trotting or jogging generates. Low-intensity work simply does not send the same signal.

WHAT HAPPENS WHEN THE STRAIN THRESHOLD IS MET

Cortical thickness increases. Cross-sectional geometry expands — the bone widens and stiffens in the load-bearing directions. Structural stiffness improves — the bone resists bending more effectively. Load tolerance rises — the bone absorbs more force before approaching its fatigue limit.

This is not optional biology. The horse that receives properly dosed high-speed work develops the bone architecture that racing demands. Eliminating high-speed work to reduce injury risk can paradoxically increase structural vulnerability. The goal is never to avoid strain. The goal is to dose it correctly.

Cyclic Loading, Microdamage, and the Repair Cycle

Each stride during high-speed galloping imposes bending stress through the cannon bone, compressive forces through the condyles of the fetlock joint, and hyperextension cycles through the suspensory apparatus. This cyclic loading produces microscopic cracks within cortical and subchondral bone. Microdamage is expected and normal — it is the cellular trigger for repair. Osteoclasts resorb the weakened tissue; osteoblasts follow, laying down new collagen matrix that mineralizes into organized, stronger bone. The result, when the cycle completes, is tissue more resilient than what preceded it.

■ THE CRITICAL VULNERABILITY WINDOW

The remodeling cycle does not strengthen bone immediately. During the resorption phase — when osteoclasts have removed damaged tissue but osteoblasts have not yet completed repair — the bone is temporarily more porous and more vulnerable than before remodeling began. The bone formation phase alone can take one to three months to complete fully. Continued high-speed loading before a remodeling cycle consolidates does not simply slow adaptation — it can actively weaken the structure. The microcracks signaling repair are still present, but the repair cycle has been interrupted before finishing. Research examining Thoroughbreds in active race training versus those at rest found that horses in continuous training showed significantly less bone resorption activity at heavily loaded sites — not a sign of health, but a sign that cyclic loading suppresses the osteoclast recruitment that initiates repair. Periods of reduced loading are necessary to allow that suppressed remodeling to occur.

Modeling vs. Remodeling in the Two-Year-Old

These two processes are related but distinct. Understanding the difference is essential to understanding why the two-year-old year is structurally irreplaceable.

MODELING	REMODELING
Active construction — changing bone mass, shape, and geometry in response to new loading demands	Ongoing maintenance — replacing fatigued or damaged tissue with fresh, organized bone
Dominant in young horses — peaks during the two-year-old year	Active throughout life — rate and repair quality change with age
Requires sufficient strain magnitude — jogging does not activate it	Requires time and reduced loading — suppressed during continuous hard training
Think: framing the house	Think: inspecting and repairing the frame

Two-year-olds possess heightened modeling responsiveness. Exercised young horses develop up to 28% greater cortical bone thickness in the cannon bone compared to unexercised controls. This window of enhanced responsiveness gradually narrows as the horse matures. The two-year-old season therefore represents a dual opportunity: build structure aggressively enough to stimulate modeling, while sequencing exposure carefully enough to allow remodeling consolidation between training blocks.

Load Magnitude vs. Load Density

Magnitude refers to the intensity of a single effort — how fast, how far, how much force per limb. A properly executed five-furlong breeze at twelve seconds per furlong represents a defined magnitude of strain. On an adequately rested skeleton, that single effort stimulates adaptation.

Density refers to how frequently high-magnitude efforts are stacked within a given time window. The same five-furlong breeze, repeated before the previous remodeling cycle has consolidated, does not double the adaptation. It adds fatigue to a structure that has not yet finished repairing from the last effort.

The stopwatch measures magnitude.
The calendar reveals density.
Both determine outcome.

A Practical Marker: The Bucked Shin Problem

Approximately 70% of young Thoroughbreds in race training develop bucked shins, and roughly 12% of those progress to stress fractures at the dorsal cannon bone. Bucked shins are a measurable signal that the balance between strain and consolidation has been lost. Short, early speed work — one furlong at approximately 14 seconds — introduced progressively before full training volumes begin has been shown to increase bone mineral content and reduce bucked shin incidence. This is the fast-and-light principle in practice: targeted strain, early, before density accumulates.

Research Foundation — Chapter 2

Strain Magnitude in the Cannon Bone: Nunamaker, D.M. et al. (1990). J Orthopaedic Research, 8(4), 604–611. Peak compressive strains of 4,000–5,000 microstrain at gallop — the threshold driving cortical modeling.

Remodeling Suppression During Training: Whitton, R.C. et al. (2014). Vet J 202(3). PubMed 25296852. LC: 0.24/mm (training) vs. 0.85/mm (resting) — threefold difference, P<0.001.

Cortical Adaptation in Exercised Horses: Firth, E.C. (2006). J Anatomy 208(4):513–526. 28% greater cortical thickness in exercised vs. non-exercised horses.

Bucked Shin Incidence: Boston & Nunamaker (2000). Am J Vet Res 61(6):602. ~70% incidence; ~12% progress to stress fracture.

2025 Subchondral Bone Study: Ciamillo, S.A. et al. (2025). Equine Vet J. DOI:10.1111/evj.14524.

CHAPTER THREE

Fatigue Accumulation and Load Density

> ***Fatigue does not announce itself. It accumulates quietly — one stride at a time — until the structure can no longer absorb what is asked of it.***

The Fatigue Fracture Mechanism

The most dangerous injuries in Thoroughbred racing are rarely caused by a single catastrophic event. They are caused by the final load placed on a structure that had been quietly weakening for weeks. Research has found that approximately 78% of racing fatalities and career-ending injuries are classified as fatigue-related — developing over time through cumulative damage rather than a single acute event.

The cannon bone condyle — the primary loading site at the fetlock joint — is the most common location for fatigue fracture in the Thoroughbred. Each racing stride produces hyperextension of the fetlock and compressive loading through the condyles. This cumulative loading across a season of training and racing is what drives the fatigue accumulation the KTM program is designed to manage.

The 60-DayThreshold

Research has identified a critical density threshold at approximately 60days of concentrated high-speed When horses accumulate high-speed exercise in a compressed eight-week window without adequate consolidation intervals, injury risk escalates significantly. This finding underpins the KTM Program's phase structure and is the biological rationale for the foundation period before any timed work is introduced.

FATIGUE COST AND CONSOLIDATION REQUIREMENTS BY EVENT TYPE

WITHDRAWAL EVENT	ESTIMATED FATIGUE COST	CONSOLIDATION REQUIRED
Single timed breeze (any distance)	Moderate microdamage at condyle and MC3; greater at longer distances	7–10 days before next timed work
Single race start (any distance)	~9.3% of subchondral fatigue life at primary loading site	4 weeks minimum before next race
Two race starts within 2 weeks	**~18.6% fatigue life — remodeling suppressed; net deficit accumulates**	**Extended reset required; monitoring essential**
Full 6-week Active Reset	**Remodeling suppression resolves; full repair cycle completes**	**Account returns to structural surplus**

Race start fatigue: Morrice-West et al. (2022) Sci Rep 12:11528.

Why Density Matters More Than Any Single Effort

It is not the hardest single workout that determines structural outcome. It is the rate at which withdrawals are stacked against a remodeling cycle that has not yet completed. When high-speed work is compressed into too short a window — before the remodeling triggered by the previous work has consolidated — fatigue accumulates faster than the remodeling cycle can resolve it, and the structural account moves into deficit without any visible warning.

This is why the KTM Program tracks cumulative high-speed exercise volume in rolling eight-week windows rather than evaluating individual workouts in isolation. A single number — total furlongs at race pace in the past 56 days — is a more reliable predictor of structural risk than any single workout time.

Fatigue and the Filly

Research has consistently found that fillies accumulate fatigue at a faster rate than colts under equivalent training loads. The difference is mechanical, not anatomical: fillies take more strides per race than colts at the same speed, meaning the same absolute training load produces more loading cycles — and therefore more accumulated fatigue — per race. A training program calibrated for colt thresholds systematically overloads fillies. The KTM Model applies a conservative density standard to all horses regardless of sex — appropriately calibrated for the more vulnerable population.

The Fatigue Debt That Crosses the Calendar

Unresolved fatigue does not reset when the horse turns three. It enters the three-year-old season as a preexisting deficit — bone asked to absorb greater loading at longer distances with less structural margin than a properly managed horse. Each race start at three draws from a reserve that was partially depleted before the season began.

THE COMPOUNDING EFFECT OF CARRIED FATIGUE

Morrice-West et al. (2022) estimated that each race start consumes approximately 9.3% of subchondral bone fatigue life at the primary loading site. A horse completing four two-year-old race starts without a subsequent six-week structural reset arrives at its three-year-old season having already consumed approximately 37% of fetlock fatigue tolerance — not from three-year-old racing, but from unresolved two-year-old racing that was never allowed to consolidate. That horse then faces longer distances, higher competition, and greater cumulative loading with a structural account that began the season already in deficit.

Research Foundation — Chapter 3

78% of Racing Fatalities Are Fatigue-Related

Stover, S.M. (2003). The epidemiology of Thoroughbred racehorse injuries. Clinical Techniques in Equine Practice, 2(4), 312–322.

~9.3% Fatigue Life Per Race Start

Morrice-West, A.V., Whitton, R.C. et al. (2022). Scientific Reports, 12, 11528. PMC9262984.

60-Day Load Window and MSI Risk

Crawford, K.L. et al. (2021). Animals, 11(1), 270. PMC7910838.

Sex Differences in Fatigue Accumulation

Morrice-West, A.V., Whitton, R.C. et al. (2022). Scientific Reports, 12, 11528. PMC9262984.

CHAPTER FOUR

Workload Pattern vs. Stopwatch Time

The stopwatch tells you how fast the horse ran. The calendar tells you whether it was safe to ask.

Half the Equation

The racing industry has long organized its evaluation of workouts around time. Twelve seconds per furlong. Eleven and four. Under eleven. These figures describe the magnitude of a single effort. But they describe only half of the structural equation. The other half is pattern: how many efforts, how close together, and what was the cumulative density of the training block in which that effort occurred.

The 35-Furlong Threshold

One of the most significant findings in equine sports medicine research is the relationship between cumulative high-speed exercise volume and fatal injury risk. Estberg et al. (1996) found that horses accumulating more than 35 furlongs of exercise faster than 14.4 seconds per furlong in the 60days prior to a race had a 3.9-fold higher risk of fatal musculoskeletal injury compared to horses below that threshold. This is not a speed finding — horses in both groups ran at similar speeds. The difference was volume: the pattern of loading across the preceding two months.

LOAD DENSITY THRESHOLDS — SAFE / ALERT / DANGER ZONES

METRIC	SAFE ZONE	ALERT ZONE	DANGER ZONE
Furlongs at race pace (prior 60days)	**< 20 furlongs**	**20–34 furlongs**	**> 35 furlongs (3.9× fatal risk)**
Race starts in 4-week window	**1 start with 4-week spacing**	**1 start with 3-week spacing**	**2+ starts in 2-week window**
Days between timed breezes	**7+ days**	**5–6 days**	**< 4 days**
Breeze distance escalation step	**Same distance repeated first**	**1-furlong step-up after repetition**	**2+ furlong jump or no repetition**

35-furlong threshold: Estberg et al. (1996) Am J Vet Res 57(11):1549-1555.

The Long Slow Distance Trap

A common misconception is that replacing high-speed work with increased slow distance work reduces structural risk. Research does not support this. Long slow distance work builds aerobic base — which is valuable — but it does not build the structural geometry that high-speed loading uniquely drives. A horse conditioned exclusively on slow work may arrive at its first race with a well-developed cardiovascular system and an under-adapted skeleton. The first race becomes the first high-magnitude load event the bone has ever experienced.

Pattern Is the Variable the Trainer Controls

The speed of a breeze is largely a function of the horse's talent and the trainer's target. The pattern of breezes — their frequency, their distance progression, their relationship to rest intervals — is entirely within the trainer's control. This is the variable that research has consistently identified as the primary modifiable risk factor for musculoskeletal injury. Not speed. Pattern.

The KTM Response: Track the Pattern, Not Just the Time

The KTM Model Program requires tracking cumulative HSE volume in rolling eight-week windows alongside individual workout times. The safe zone threshold — below 20 furlongs of high-speed work in any 60-day period during the foundation and escalation phases — is derived directly from the Estberg threshold. Crossing into the alert zone (20-34 furlongs) triggers heightened structural monitoring. Approaching the danger zone (35+ furlongs) requires immediate density reduction regardless of how sharp the horse's recent works have been.

Research Foundation — Chapter 4

35-Furlong Threshold and Fatal Injury Risk

Estberg, L., Stover, S.M., Gardner, I.A. et al. (1996). American Journal of Veterinary Research, 57(11), 1549-1555. PubMed 8915427. 3.9-fold increase in fatal MSI risk for horses exceeding 35 furlongs faster than 14.4 sec/f in the 60 days prior to a race.

Cumulative Load Volume as Primary Modifiable Risk Factor

Crawford, K.L. et al. (2021). PMC7910838. Morrice-West, A.V. et al. (2022). PMC9262984. Wong, A.S.M. et al. (2023). Front Vet Sci 10:1260554.

High-Speed Exercise as Necessary for Structural Adaptation

Goodship, A.E. et al. (1991). Equine Veterinary Journal, 23(3), 202–206. PubMed 1553435. Near-maximal speed required to produce periosteal bone formation.

CHAPTER FIVE

The KTM Structural Framework

Five principles. One integrated architecture. Each reinforces the others — none stands alone.

The preceding four chapters established the biological foundation: bone responds to strain, adapts when loaded progressively, degrades when loaded without consolidation, and accumulates fatigue that compounds across seasons. This chapter translates that biology into the five structural principles that govern every decision in the KTM Model Program.

PRINCIPLE 1: PROGRESSIVE ESCALATION

Distance and speed are introduced in measurable increments following a structured sequence. The program opens with two single introductory works — a 1/8 mile breeze followed by a 1/4 mile breeze — each done once as stepping stones to reach the distance where the confirmation rule begins. From the 3/8 mile onward, no distance is advanced from until it has been worked twice. The first work at any new distance is a controlled, easier effort — a structural introduction. The second work is the confirmation at the 12-second-per-furlong standard. Only after that confirmation does the program advance to the next distance. The full sequence is: 1/8 (once) → 1/4 (once) → 3/8 → 3/8 confirmed → 1/2 → 1/2 confirmed → 5/8 → 5/8 confirmed → 1/2 step back.

PRINCIPLE 2: DENSITY ACCOUNTING

Total high-speed exercise furlongs are tracked in rolling 60-daywindows. The safe zone ceiling — below 20 furlongs per 60days during foundation and escalation phases — is derived from the Estberg threshold (1996). Crossing into the alert zone triggers structural monitoring. The danger zone (35+ furlongs) within a 60-day window requires immediate density reduction.

PRINCIPLE 3: REPETITION BEFORE ESCALATION

The 1/8 mile and 1/4 mile opening works are each performed once — they are introductory works, not subject to the repetition rule. Beginning at the 3/8 mile, every distance must be worked twice before the program advances. The first work at any new distance from 3/8 onward is a controlled introduction. The second work is the confirmation. No distance from 3/8 through 5/8 is attempted only once before advancing. Escalation from 3/8 onward is earned through repetition — not assumed from a single clean workout.

PRINCIPLE 4: TWO SIX-WEEK STRUCTURAL RESETS

Twice within the two-year-old year, high-speed work is fully suspended. This is not rest from fatigue — it is a biological reset. During six weeks of rest, remodeling suppression resolves, the full repair cycle activates, and cortical consolidation occurs. The reset is the deposit that funds the next competitive phase.

PRINCIPLE 5: RACE SPACING DISCIPLINE

Race starts are treated as high-magnitude load events consuming approximately 9.3% of subchondral fatigue life. A four-week minimum between race starts is the KTM standard. Three-week spacing is a monitored exception. Two-week spacing is avoided except in extraordinary circumstances with confirmed structural soundness.

A NOTE ON PRINCIPLES 1 AND 3 — THE MAP AND THE RULE

Principles 1 and 3 work together but serve different purposes and it is worth being clear about the distinction. Principle 1 — Progressive Escalation — describes the map: the specific sequence of distances the program follows, starting with the single introductory 1/8 and 1/4 mile works and then building through 3/8, 1/2, and 5/8. It tells you where the program goes and in what order. Principle 3 — Repetition Before Escalation — describes the rule that governs the map from 3/8 onward: no distance advances until it has been worked twice. The first work is a structural introduction at a controlled pace. The second is the confirmation at the 12-second-per-furlong standard. It tells you why you cannot simply follow the sequence faster. In plain terms: Principle 1 is the road. Principle 3 is the speed limit on that road — and it applies from the 3/8 mile onward.

How the Five Principles Work Together

The five principles are interlocking constraints that reinforce each other. Progressive escalation controls the magnitude of individual loading events. Density accounting controls the frequency at which those events are stacked. Repetition before escalation ensures each level is confirmed before the next is attempted. The two reset blocks provide the consolidation windows that allow the preceding work to integrate. Race spacing discipline prevents competition from consuming structural margin faster than training can replenish it.

Remove any one of these principles and the architecture develops a structural weakness. Escalate without repetition and you advance without confirmation. Stack density without accounting and you approach the fatigue threshold without knowing it. Skip the reset blocks and you enter the competition phase with unresolved remodeling debt. Compress race spacing and you consume fatigue life faster than training can recover it.

What the Framework Does Not Restrict

The KTM framework does not restrict competitive ambition. It does not argue against the twelve-second furlong, against stakes-race targets, or against racing two-year-olds at all. What it restricts is compression — the tendency to stack escalation faster than the remodeling cycle can accommodate. Competitive horses are produced within this framework every season. The difference is that they arrive at three and four with structural margin intact rather than depleted.

The framework does not slow the horse. It sequences the load. The horse that benefits most from architecture is not the fragile horse. It is the talented horse — because talent is worth protecting.

Research Foundation — Chapter 5

The Five Principles — Research Foundations

P1 Progressive Escalation: Crawford et al. (2021) PMC7910838. P2 Density Accounting: Estberg et al. (1996) JAVMA 208:92; Morrice-West et al. (2022). P3 Repetition: Wong et al. (2023) Front Vet Sci 10:1260554. P4 Reset Blocks: Whitton et al. (2014) PubMed 25296852. P5 Race Spacing: Morrice-West et al. (2022); Burns (2021) PMC7916178.

PET Scan Evidence for Rest Block Efficacy

65% of horses with fetlock lesions showed improvement after 12-week rest in PET imaging studies. Referenced in Burns (2021) PMC7916178 and Equine Vet J review.

CHAPTER SIX

The Traditional Twelve-Second Model

The twelve-second furlong is not the problem. Compression is the problem. The KTM Model preserves the standard — and sequences it correctly.

What the Traditional Model Gets Right

The twelve-second furlong is the heartbeat of American Thoroughbred training. It is the speed at which young horses learn to travel, the standard by which readiness is measured, and the cultural currency of the backstretch. The traditional model's instinct is correct at the level of individual effort. High-speed work is necessary for cortical adaptation. The twelve-second standard produces the strain magnitude required to drive structural modeling in young bone. The KTM Model does not ask trainers to abandon this.

Where Compression Creates Risk

The problem arises not in the speed of individual breezes but in how quickly the program escalates from first exposure to competition distance. Research has found that approximately 55% of two-year-olds required veterinary attention for musculoskeletal issues in their first racing season — a figure that reflects not the inherent danger of two-year-old racing but the consequences of compressed escalation in the weeks preceding it.

TRADITIONAL MODEL vs. KTM MODEL — KEY DIFFERENCES

DIMENSION	TRADITIONAL MODEL	KTM MODEL
Speed standard	12 sec/f	12 sec/f (unchanged)
Foundation period before first timed work	Variable; often 4–6 weeks	Foundation period
Breeze distance escalation	As fast as horse shows readiness	Repetition before escalation; Distance confirmed twice
Reset blocks	Not structured; lay-ups reactive to injury	Two planned 6-week resets; Proactive
Race spacing	3–4 week spacing common	4-week target; 3-week monitored exception
Cumulative HSE tracking	Rarely tracked formally	Tracked in 60-dayrolling windows

The Cultural Bridge

The most important thing the KTM Model does culturally is preserve the metric everyone in the industry already understands — the twelve-second furlong — while changing the framework around which it is deployed. Trainers do not need to relearn how to train horses. They need to relearn how to sequence what they already do. The speed stays the same. The architecture changes.

Research has found that each additional furlong per day of high-speed exercise was associated with a 73% increase in musculoskeletal injury odds. This is a density finding, not a speed finding. It tells trainers not to run slower — it tells them to run the same speed less often in compressed windows. That is a scheduling change, not a philosophical one.

THE CULTURAL BRIDGE IN PRACTICE

The twelve-second furlong remains. Competition remains. Ambition remains. What changes is compression tolerance. The KTM Model introduces biological respect into cultural rhythm. Trainers do not need to relearn how to train horses — they need to relearn how to sequence what they already do. The speed stays the same. The architecture changes.

Research Foundation — Chapter 6

55% Veterinary Attention Rate in First Racing Season

Referenced in Crawford et al. (2021) PMC7910838 and Burns (2021) PMC7916178. Approximately 55% of two-year-olds required veterinary attention for musculoskeletal issues in their first racing season — a figure that reflects the consequences of compressed escalation, not the inherent danger of two-year-old racing.

73% Increase in MSI Odds Per Additional Furlong Per Day

Morrice-West, A.V. et al. (2020). Associations between pre-race training history and race-day MSI in Thoroughbreds. PMC7696103. Each additional furlong per day of high-speed exercise was associated with a 73% increase in musculoskeletal injury odds — a density finding, not a speed finding.

12-Second Standard and Modeling Threshold

Nunamaker et al. (1990) J Orthop Res 8(4):604. Speed faster than 14.4 sec/f required to exceed the strain threshold that drives cortical modeling. The twelve-second furlong standard correctly meets this threshold — the issue is not the speed but the density and sequencing of its application.

CHAPTER SEVEN

Conservative Load Introduction Models

Too little loading produces an under-adapted skeleton. Too much loading produces a fatigued one. Research defines the window between them.

The Non-Linear Relationship

The instinct to protect young horses by limiting speed is understandable and partially correct. The instinct to accelerate training for early competitive results is equally understandable and partially correct. Research has clarified the relationship: injury risk follows a non-linear curve. It is elevated at low high-speed exercise volumes and elevated again at high volumes. The lowest injury risk sits in the middle range.

Wong et al. (2023) found that trainers who preferred higher high-speed exercise volumes had horses with lower musculoskeletal injury risk compared to trainers who preferred lower volumes — up to a point. Horses with no high-speed work were at elevated risk. Horses with excessive high-speed work were at elevated risk. Horses with intermediate, appropriately spaced high-speed work had the lowest risk profile.

HIGH-SPEED EXERCISE VOLUME AND INJURY RISK — THE NON-LINEAR RELATIONSHIP

HSE VOLUME ZONE	STRUCTURAL EFFECT	INJURY RISK PROFILE
Very low (< 8 furlongs/60 days)	Insufficient strain — cortical modeling does not activate	ELEVATED Skeleton under-adapted
Moderate (8–20 furlongs/60 days)	Progressive adaptation — modeling exceeds microdamage accumulation	LOWEST Optimal adaptation range
High (20–34 furlongs/60 days)	Adaptation slowing — microdamage accumulates faster than remodeling	ALERT Monitor closely
Excessive (> 35 furlongs/60 days)	Fatigue dominates — remodeling suppressed; structural margin erodes	DANGER 3.9× fatal injury risk

Non-linear relationship: Wong et al. (2023) Front Vet Sci 10:1260554.

The Modeling Threshold: Why Slow Work Is Not Enough

Research by Goodship et al. (1991) found that only near-maximal speed exercise produced periosteal bone formation in young horses. Work at lower intensities — even sustained long-distance slow work — did not produce the same cortical response. A conservative load introduction program must therefore include genuine high-speed work from its earliest phases — but at low volume, with adequate spacing, before escalating.

10–14 Week Foundation Window

Crawford et al. (2021) found that horses in their first 10–14 weeks of training had an odds ratio of 8.05 for musculoskeletal injury (95% CI 1.92, 33.76; p=0.004). This is the highest-risk period in the entire two-year-old career. The KTM Phase I extends past this window before any timed work is introduced — not as caution, but because the biology demands it.

Research Foundation — Chapter 7

Non-Linear HSE Volume and Injury Risk

Wong, A.S.M. et al. (2023). Front Vet Sci, 10, 1260554.

Near-Maximal Speed Required for Bone Formation

Goodship, A.E. et al. (1991). Equine Vet J, 23(3), 202–206. PubMed 1553435.

10–14 Week MSI Risk Window

Crawford, K.L. et al. (2021). Animals 11(1):270. PMC7910838. OR 8.05, p=0.004.

CHAPTER EIGHT

KTM Program Architecture

Architecture is the difference between a program that produces sound horses and one that produces injured ones from the same raw material.

The preceding seven chapters established the biological and research foundation of the KTM Model. This chapter presents the architecture that translates that foundation into a practical six-phase program governing the entire two-year-old season — from first training day to the transition into the three-year-old year.

The Six-Phase Structure

The KTM two-year-old season is organized into six sequential phases. Each phase has a defined structural objective, a prescribed load profile, and specific exit criteria before transition to the next phase. No phase is optional and none can be compressed without explicit biological justification and documented structural monitoring.

KTM SIX-PHASE ARCHITECTURE — COMPLETE OVERVIEW

PHASE	WEEKS	LABEL	STRUCTURAL OBJECTIVE	KEY CONSTRAINT
I	1–10	Foundation	Build aerobic base; establishes structural foundation before first timed work	No timed breezes; slow gallop only
II	11–19	Breeze Progression	First high-speed exposure; establish modeling threshold signal; build to competition distance	1/8 (once) → 1/4 (once) → 3/8 → 3/8 → 1/2 → 1/2 → 5/8 → 5/8 → step back
III	20–25	Active Reset I	Allow suppressed remodeling to complete full repair cycle	No timed breezes; 6 weeks minimum
IV	26–28	Competition Prep	Return to work after reset; confirm competition readiness through graduated breezes	Single easy 3/8 → 1/2 → 5/8 → race entry; 4-week spacing standard begins
V	29–40	Competition	Race starts with structural accounting; 4-week spacing standard	Race + gallop-out counts toward HSE total
VI	41–46	Active Reset II	Clear accumulated competition fatigue; prepare 3YO season	No timed breezes; 6 weeks minimum

Phase Dependency: Why Order Cannot Be Reversed

Each phase depends on the completion of the phase before it. Phase II can only begin after Phase I has established the structural foundation. Phase III can only begin after Phase II has confirmed structural tolerance at 3f and 4f through repetition. Phase IV can only begin after Phase III has provided six full weeks for remodeling consolidation. Phase V can only begin after Phase IV has confirmed competition readiness. Phase VI must follow Phase V regardless of competitive calendar pressure.

The Six-Week Reset: Biological Mechanism

The six-week duration of each reset block is not arbitrary. Whitton et al. (2014) documented that remodeling suppression in continuously trained horses resolves within six weeks of rest — eroded bone surface returns to normal levels, and the labeling coefficient rises from 0.24 per mm in training horses to 0.85 per mm in resting horses. Six weeks is the minimum time required for the remodeling cycle to progress from suppression to completion.

THE PHASE III AND VI RESET — WHAT HAPPENS BIOLOGICALLY

Weeks 1–2: Removal of high-speed loading allows osteoclast recruitment to begin at sites where remodeling was suppressed. Resorption phase initiates. Week 2–3: Resorption phase peaks. Bone is temporarily more porous than before rest began — the return-to-work vulnerability window. No high-speed work. Weeks 3–5: Osteoblasts follow osteoclasts. New bone matrix is laid down and begins to mineralize. Porosity decreases. Week 6: Mineralization consolidates. Bone formation markers return to baseline. Structural account is replenished. Horse is ready for graduated return to work.

The program is not conservative. It is precise. Every phase has a biological reason. Every constraint has a research anchor. Every decision is architectural — not cautious.

Research Foundation — Chapter 8

10–14 Week MSI Risk Window

Crawford, K.L. et al. (2021). Animals 11(1):270. PMC7910838. OR 8.05 (p=0.004).

Six-Week Reset Biology

Whitton, R.C. et al. (2014). Vet J 202(3). PubMed 25296852. LC: 0.24/mm (training) vs. 0.85/mm (resting) — threefold difference, P<0.001.

Race Start as Load Event

Morrice-West, A.V. et al. (2022). Sci Rep 12:11528. PMC9262984. ~9.3% per start.

Phase V Race Spacing

Burns, T.A. (2021). PMC7916178. Morrice-West et al. (2022). Sci Rep 12:11528.

CHAPTER NINE

The First Six-Week Remodeling Block

The first reset is not a pause in the program. It is the program — doing exactly what it was designed to do.

Why Suppressed Remodeling Must Be Released

Phase IV of the KTM Model is the first six-week Active Structural Reset. It arrives after the breeze progression has taken the horse from 1/8 mile through to two confirmed 5/8 mile works and a deliberate step-back to 1/2 — and before the horse enters competition for the first time. Whitton et al. (2014) provided the critical data: horses in continuous race training showed the labeling coefficient — a direct measure of active bone formation — at 0.24 per mm versus 0.85 per mm in resting horses. A threefold difference. Continuous training suppresses the osteoclast recruitment that initiates repair. The microdamage accumulated during the escalation phase has not been repaired. It is waiting for the signal that rest provides.

The key biological distinction is this: it is HIGH-SPEED CYCLIC LOADING that suppresses remodeling — not movement itself. Tack walking, jogging and easy galloping below the 15-second-per-furlong threshold do not generate the strain magnitude that inhibits osteoclast recruitment. The horse can and should remain in daily light work throughout Phase IV. This keeps the horse mentally settled, maintains cardiovascular base, preserves muscle condition, and — critically — keeps the horse in regular visible work that satisfies modern veterinary oversight requirements.

THE SIX-WEEK RESET — WHAT IT IS AND IS NOT	
A biological requirement — six weeks minimum for the full remodeling cycle to complete	A sign of injury or weakness — structurally sound horses reset on schedule
A structural deposit — remodeling suppression resolves, repair activates, bone consolidates	Lost time — the horse returns structurally stronger than it left
A proactive protocol — planned at program design, not triggered by soreness	A reactive lay-up — reactive lay-ups respond to failure; resets prevent it
Part of competitive preparation — the horse enters Phase V structurally sound	An interruption to training — it is training, expressed as biological consolidation

The Porosity Window — Understanding Mid-Reset Vulnerability

During the resorption phase of remodeling — approximately weeks two and three of rest — bone is temporarily more porous than before rest began. Osteoclasts have removed fatigued tissue but osteoblasts have not yet completed the replacement. This is the most vulnerable point in the rest period. It is why abbreviated rest blocks — three or four weeks — can be more dangerous than continuous training. The horse enters the resorption phase, becomes temporarily weaker, and is returned to high-speed work before the repair phase has completed. A partial reset is worse than no reset at all.

Return to Work: The Ten-Day Hazard Period

Carrier et al. (1998) identified a significant association between long periods without high-speed work and fracture risk upon return to training — specifically in the first ten days after return. This finding is not a contradiction of the reset block rationale. It is an operational instruction for how to execute the return.

PHASE IV RETURN-TO-WORK PROTOCOL — WEEK BY WEEK

RETURN WEEK	WORK TARGET	RATIONALE
Week 1 (highest risk)	No timed work. Daily tack walk, jog and easy gallop only.	Porosity resolving; new matrix still maturing. Do not interrupt consolidation.
Week 2	Single easy 3/8 at :37+.	First mechanical signal back — easy controlled pace. Not a performance test.
Week 3	1/2 mile at :49+.	First work back at 1/2 — controlled.
Week 4	5/8 mile at 1:01+.	First work back at competition distance — controlled.
Week 5+	Normal Phase V prep resumes.	Full competition preparation resumes under density accounting rules.

Ten-day hazard period: Carrier et al. (1998) JAVMA 212(10):1582.

A horse that enters its first race after a completed six-week reset is a structurally different horse from one that races straight out of an escalation phase. Both may look sharp in the morning. Only one has the structural account to prove it.

Research Foundation — Chapter 9

Remodeling Suppression in Training — LC Data

Whitton, R.C. et al. (2014). Vet J 202(3). PubMed 25296852. LC 0.24/mm (training) vs. 0.85/mm (resting) — threefold difference, P<0.001.

Return-to-Work Vulnerability — Ten-Day Hazard Period

Carrier, T.K., Estberg, L., Stover, S.M. et al. (1998). JAVMA, 212(10), 1582.

Porosity During Resorption Phase

Martig, S. et al. (2014). Vet J 202(3), 566–575. PubMed 24528139.

80% Fatigue Injury Prevalence

Whitton, R.C. (2018). Subchondral bone fatigue and remodelling. PubMed 29660153.

CHAPTER TEN

The Second Structural Reset

The second reset carries a different weight — and a different purpose. It is the deposit that funds the three-year-old year.

Why Fatigue Accumulates Across the Competition Phase

Phase VI of the KTM Model — the second six-week structural reset — arrives after the competition phase has concluded. Where Phase IV prepared the horse for its first race starts, Phase VI prepares it for an entire subsequent season. It is not recovery from injury. It is a deliberate architectural decision to carry zero fatigue debt across the calendar boundary into the three-year-old year.

The competition phase is the highest-density period in the two-year-old season. Each race start consumes approximately 9.3% of subchondral bone fatigue life. Between-race training maintains baseline HSE volume. Partial consolidation occurs between race starts with four-week spacing — but it does not produce the complete remodeling cycle that six weeks of rest provides.

CUMULATIVE FATIGUE STATUS AT END OF PHASE V — FOUR SCENARIOS

COMPETITION PHASE SCENARIO	FATIGUE STATUS AT PHASE END
3 race starts, 4-week spacing, Phase VI reset completed	~27.9% fatigue life consumed from races. Full reset resolves residual. Horse enters 3YO at full structural margin.
4 race starts, 3-week spacing, Phase VI reset completed	~37.2% consumed. Reset resolves residual accumulation. 3YO entry structurally sound.
5 race starts, 2–3 week spacing, NO Phase VI reset	~46.5% consumed with no resolution. Horse enters 3YO with nearly half fatigue life already spent.
6 race starts, compressed spacing, NO reset	~55.8% consumed. 3YO season begins from structural deficit. Wong et al. (2023) six-race warning signal breached.

~9.3% fatigue life per race start: Morrice-West et al. (2022) Sci Rep 12:11528.

Four Objectives of the Second Reset

Phase VI serves four distinct objectives simultaneously — none of which can be achieved by simply reducing training intensity without fully suspending high-speed work.

FOUR OBJECTIVES OF THE PHASE VI RESET

OBJECTIVE	BIOLOGICAL MECHANISM	CONSEQUENCE IF SKIPPED
Clear competition fatigue from subchondral bone	Full remodeling cycle activates without high-speed interference	Unresolved microdamage carries into 3YO year
Allow skeletal muscle transition to complete	136 DEGs associated with muscle assembly upregulate during 2-to-3-year transition	Muscle structural development competes with training demands
Restore structural margin to surplus	Remodeling markers return to resting levels; labeling coefficient rises threefold	Horse begins 3YO season with depleted structural account
Prevent fatigue carryover into 4YO year	Each unresolved race start compounds across seasons at ~9.3% per start	Three-year-old racing amplifies 2YO deficit; 4YO dividend diminished

Momentum vs. Architecture at the End of the Season

The pressure against Phase VI is momentum. A horse competing well, whose connections are excited, whose form is sharp — this horse is difficult to put away. The calendar shows attractive race opportunities. The structural biology does not appear on the entry form.

The KTM response is direct: the momentum that makes Phase VI feel premature is precisely the evidence that the horse has been doing its job correctly. A horse in form is drawing from its structural account at a healthy rate. That account needs replenishment before the next campaign begins. Momentum is not a substitute for consolidation.

The horse that completes Phase VI is not a horse that was pulled up short. It is a horse whose connections understood that the four-year-old dividend is funded by decisions made at the end of the two-year-old year. The reset is not the end of the season. It is the beginning of the next one.

Research Foundation — Chapter 10

Fatigue Life Per Race Start

Morrice-West, A.V. et al. (2022). Sci Rep 12:11528. PMC9262984. ~9.3% per start.

Six-Race Warning Signal

Wong, A.S.M. et al. (2023). EVJ 55:194. PMC10084173. 18% MSI risk increase per 0.1 m/s speed decrease; 11% per 10 cm stride decrease.

Career Longevity — 2YO Racing Association

Tanner, J.C. et al. (2013). EVJ 45:20. Velie, B.D. et al. (2013). EVJ 45:410. ~3 more career starts from age 3 onward for horses racing as 2-year-olds.

Skeletal Muscle Transition 2YO to 3YO

2024 JAVMA transcriptome analysis: 136 differentially expressed genes in skeletal muscle between 2YO and 3YO at rest. Key pathways: muscle assembly, growth factor binding, ECM-receptor interaction.

CHAPTER ELEVEN

The Full Two-Year-Old Conditioning Calendar

The calendar is not the constraint. The biology is the constraint. The calendar is the tool we use to respect it.

This chapter presents the full 46-week operational roadmap — a week-by-week structure that translates every principle into practice. It is an architectural framework within which individual horse responses guide specific decisions, not a rigid schedule that ignores variation.

THE 46-WEEK SEASONAL ARC

I Wks 1–10 FOUNDATION	II Wks 11–19 BREEZES	Reset I Wks 20–25	Comp Prep Wks 26–28	V Wks 29–40 COMPETITION	Reset II Wks 41–46

PHASE I — FOUNDATION (WEEKS 1–10)

WEEK	WORK TYPE	STRUCTURAL PURPOSE
1–2	Tack walking only	Acclimation; no musculoskeletal loading stimulus
3	Light jogging begins — conditional on trainer and veterinary assessment	First introduction to jogging; below modeling threshold
4	Light jogging continues; assess progression; vet monitoring ongoing	Aerobic conditioning building; transition to galloping when horse going consistently well
5	Jogging + easy galloping — transitioning as horse settles	Aerobic base building; sub-threshold
6	Easy galloping — horse galloping consistently	Building aerobic base; preparing for 1/8 mile open gallops
7	1/8 mile open gallops begin — 1st increment; open gallop, not timed	Initiates bone remodeling; begins prevention of bucked shins
8	1/8 mile open gallops — 2nd increment	Bone mineral content increasing; easy galloping on off days
9	1/8 mile open gallops — 3rd increment	Bone responding to load; easy galloping on off days
10	1/8 mile open gallops — 4th increment; final week before first breeze	Structural base established; vet clearance before Phase II

PHASE II — BREEZE PROGRESSION (WEEKS 11–19)

WEEK	TIMED WORK	RULE APPLIED
11	1/8 mile — first breeze	Single introductory breeze. Not repeated.
12	1/4 mile — first timed work	:26–:27 controlled. Single introductory work. Not repeated.
13	3/8 mile — first exposure	:37+ controlled. Repetition rule begins. Must confirm before advancing.
14	3/8 mile — confirmed	:36. Structural tolerance confirmed. Advance to 1/2.
15	1/2 mile — first exposure	:49+ controlled. Confirm before sharpening.
16	1/2 mile — confirmed	:48. Structural tolerance confirmed. Advance to 5/8.
17	5/8 mile — first exposure	1:01+ controlled. Confirm before sharpening.
18	5/8 mile — confirmed	1:00. Competition distance confirmed. Step back next week.
19	1/2 mile — step back	:49. Deliberate density reduction before Active Reset I.

PHASE III — ACTIVE RESET I (WEEKS 20–25)

WEEK	ACTIVITY	STRUCTURAL STATUS
20–21	Tack walking, jogging and easy galloping daily	High-speed loading removed. Remodeling suppression begins to resolve.
22–23	Jogging and easy galloping daily — POROSITY WINDOW	Resorption phase peaks. No high-speed loading under any circumstance.
24–25	Easy galloping — extend distance as consolidation progresses	Bone formation phase active; consolidation occurring. Fitness maintained.

PHASE IV — COMPETITION PREP — RETURN TO WORK (WEEKS 26-28)

WEEK	TIMED WORK	RATIONALE
26	Single easy 3/8 at :37+	First mechanical signal after reset. Controlled — not a performance test.
27	1/2 mile at :49+	First work back at 1/2 — controlled.
28	5/8 mile at 1:01+	First work back at competition distance — controlled. Race entry evaluation begins.
29+	Race entry. Maintenance 1/2 at :49 between starts.	Horse structurally prepared. 4-week spacing standard between race starts.

PHASE V — COMPETITION (WEEKS 29–40)

RACE SPACING	STRUCTURAL ASSESSMENT	KTM GUIDANCE
~4 weeks (28+ days)	Partial remodeling consolidation; fatigue at manageable rate	PREFERRED — standard for all Phase V entries
~3 weeks (21–27 days)	Minimal consolidation; cumulative fatigue rises faster	USE SPARINGLY — requires clean monitoring
< 2 weeks (< 14 days)	No consolidation possible; structural deficit compounds	AVOID in 2YO season

PHASE VI — ACTIVE RESET II (WEEKS 41–46)

WEEK	ACTIVITY	3YO PREPARATION
41–42	Tack walking, jogging and easy galloping daily	Competition fatigue begins to resolve. Remodeling cycle reactivating.
43–44	Jogging and easy galloping	Remodeling cycle completing; skeletal muscle transition active.
45–46	Easy galloping — extend distance as consolidation progresses	Consolidation completing. Structural account replenished. 3YO season begins from surplus.

The full calendar is forty-six weeks. Twelve of those weeks are structural resets. Those twelve weeks are not time away from training. They are the training — expressed as biological consolidation. The horse that completes all six phases arrives at the three-year-old season with the structural margin its talent deserves.

Research Foundation — Chapter 11

Phase I Duration — 10–14 Week MSI Risk Window

Crawford et al. (2021) Animals 11(1):270. PMC7910838. OR 8.05 (p=0.004).

Phase II — Repetition Before Escalation

Wong et al. (2023) Front Vet Sci 10:1260554.

Phase IV and VI — Six-Week Reset Biology

Whitton et al. (2014) PubMed 25296852. Carrier et al. (1998) JAVMA 212(10):1582.

Phase V — Race Spacing

Morrice-West et al. (2022) Sci Rep 12:11528. Burns (2021) PMC7916178.

CHAPTER TWELVE

Breeze Sequencing and Load Tables

A table does not make training decisions. It makes the consequences of decisions visible before they become injuries.

Defining High-Speed Exercise

For KTM load accounting, high-speed exercise is any work performed at or faster than 15 seconds per furlong. This threshold encompasses all timed breezes at competitive pace, all race starts, and the gallop-out distance following each timed effort. Slow gallop sets below this threshold are aerobic work and do not count toward the 60-day HSE total.

HSE DEFINITION — THRESHOLD VALUES BY MEASUREMENT SYSTEM

MEASUREMENT	HSE THRESHOLD	NON-HSE (AEROBIC)
Seconds per furlong	≤ 15.0 sec/f	> 15.0 sec/f
Meters per second	≥ 13.3 m/s	< 13.3 m/s
Meters per minute	≥ 800 m/min	< 800 m/min
Miles per hour	≥ 30 mph	< 30 mph

A NOTE ON GALLOP-OUT PACE

Gallop-out following a breeze is expected to be a controlled deceleration, not sustained effort — well below the 15 sec/f threshold defined above. Run at this pace, gallop-out furlongs are aerobic and do not count toward the HSE total. If a gallop-out is allowed to carry speed at or faster than 15 sec/f, it must be counted toward the total. When in doubt, time it.

PHASE II — BREEZE PROGRESSION SEQUENCE (WEEKS 11–19)

WEEK	DISTANCE	TARGET TIME	RULE
11	1/8 mile	Easy — open effort	Single introductory breeze. Not repeated.
12	1/4 mile	:26–:27 controlled	Single introductory work. Not repeated.
13	3/8 mile — first exposure	:37+	Repetition rule begins. Confirm before advancing.
14	3/8 mile — confirmed	:36	Structural tolerance confirmed. Advance to 1/2.
15	1/2 mile — first exposure	:49+	Confirm before sharpening.
16	1/2 mile — confirmed	:48	Structural tolerance confirmed. Advance to 5/8.
17	5/8 mile — first exposure	1:01+	Confirm before sharpening.
18	5/8 mile — confirmed	1:00	Competition distance confirmed. Step back next week.
19	1/2 mile — step back	:49	Density reduction before Active Reset I.

ACTIVE RESET I AND RETURN SEQUENCE

WEEK	ACTIVITY	TARGET	STATUS
20–25	Active Reset I — tack walking, jogging and easy galloping.	Sub-threshold	Remodeling cycle completes
26	3/8 mile — first breeze back	:37+	Controlled return — not a performance test
27	1/2 mile	:49+	First back at 1/2 — controlled
28	5/8 mile	1:01+	Competition distance — race ready

PHASE V — RACE SPACING COMPARISON TABLE

RACE SPACING	RECOVERY DAYS	CONSOLIDATION AVAILABLE	KTM RATING
5 weeks (35+ days)	35+	Strong partial remodeling consolidation; structural markers typically normalize	IDEAL
4 weeks (28 days)	28	Adequate partial consolidation	PREFERRED
3 weeks (21 days)	21	Minimal consolidation; fatigue accumulates faster than remodeling resolves	MONITOR
2 weeks (14 days)	14	Essentially none; structural deficit compounds with each start	AVOID

MASTER HSE LOAD THRESHOLDS — QUICK REFERENCE

METRIC	SAFE ZONE	ALERT ZONE	DANGER ZONE	SOURCE
Furlongs at HSE pace (prior 60 days)	< 20 furlongs	20–34 furlongs	> 35 furlongs 3.9× fatal risk	Estberg 1996
Race starts in 4-week window	1 start, 4-week gap	1 start, 3-week gap	2+ starts in 14 days	Morrice-West 2022
Days between race and next timed work	10+ days	7–9 days	< 7 days	KTM framework
Days between timed breezes	7+ days	5–6 days	< 4 days	Crawford 2021
Breeze escalation step	1 furlong after repetition	1 furlong, no repetition	2+ furlongs in one step	Wong 2023

Safe/Alert/Danger designations are operational guidelines derived from population-level research.

Research Foundation — Chapter 12

HSE Threshold Definition

Morrice-West et al. (2020). PMC7696103. ≥15 sec/furlong definition.

35-Furlong Danger Zone

Estberg et al. (1996). JAVMA 208(1):92. PubMed 8915427.

Race Start Fatigue Accounting

Morrice-West et al. (2022). Sci Rep 12:11528.

Return-to-Work After Extended Rest

Carrier et al. (1998). JAVMA 212(10):1582.

CHAPTER THIRTEEN

Strategic Competition Planning

A race is not an opportunity independent of context. It is a load event that must fit within an architectural sequence.

The Race as a Load Event

Each race start consumes approximately 9.3% of subchondral bone fatigue life. Race conditions amplify every loading variable that governs training risk: higher speed, maximal effort, competitive interference, and often deeper ground than morning training. A race start plus a six-furlong gallop-out represents a combined HSE event of eight to ten furlongs — a full week's worth of training-phase high-speed volume in a single afternoon.

RACE SPACING COMPARISON TABLE

RACE SPACING	RECOVERY DAYS	CONSOLIDATION AVAILABLE	KTM RECOMMENDATION
5 weeks (35+ days)	35+	Strong partial remodeling consolidation; structural markers typically normalize	IDEAL for 2YO season
4 weeks (28 days)	28	Adequate partial consolidation	PREFERRED — standard target
3 weeks (21 days)	21	Minimal consolidation; fatigue accumulates faster than remodeling resolves it	USE SPARINGLY — require clean monitoring
2 weeks (14 days)	14	Essentially none; structural deficit compounds with each start	AVOID in 2YO season

The Six-Race Warning Signal

Wong et al. (2023) documented that horses sustaining musculoskeletal injuries showed a statistically measurable decline in racing speed and stride length beginning approximately six races before injury. Risk of MSI increased by 18% for each 0.1 meters per second decrease in racing speed, and by 11% for each 10 centimeter decrease in stride length across career race starts. Structural compromise is visible in performance data before it becomes a clinical injury.

THE SIX-RACE WARNING SIGNAL — WHAT TO WATCH
SPEED: A measurable decline in average racing speed across consecutive starts (18% increased MSI risk per 0.1 m/s decrease). STRIDE: Shortening of average stride length across consecutive starts (11% increased MSI risk per 10 cm decrease). RECOVERY: Longer time to return to normal digital pulse, thermal symmetry, and limb filling after each race compared to earlier starts. BREEZE QUALITY: Post-race training breezes progressively flatter compared to pre-race breezes at the same distance and surface. If two or more of these signals appear across three consecutive starts, structural reassessment is required before the next entry.

Race Selection Framework

Race selection in the KTM program is governed by five criteria evaluated in sequence. A race that fails any criterion is not entered, regardless of purse size or competitive opportunity.

RACE SELECTION DECISION FRAMEWORK — FIVE CRITERIA

CRITERION	QUESTION TO ANSWER	REQUIRED STATUS BEFORE ENTRY
1. Structural Recovery	Has adequate time elapsed since last start for partial remodeling consolidation?	Minimum 28 days (21 days only with clean monitoring confirmation)
2. Structural Monitoring	Are all five monitoring indicators within normal range?	All five indicators normal. Any anomaly requires investigation before entry.
3. Distance Appropriateness	Has the horse been prepared for this distance in training?	Race distance must not exceed longest timed work by more than 1 furlong.
4. 60-Day HSE Total	Will this race entry push the rolling 60-day HSE total into the danger zone?	Race + gallop-out must not push 60-day total above 35 furlongs.
5. Travel and Surface	Do travel demands and surface conditions add structural risk?	Long-haul travel within 5 days of race requires additional recovery monitoring.

Between-Race Structural Monitoring

The five-indicator monitoring protocol is applied at 48 hours post-race, at seven days post-race, and at the point of training resumption. Any indicator outside normal range extends the consolidation period and delays the next training decision.

BETWEEN-RACE STRUCTURAL MONITORING — FIVE INDICATORS

INDICATOR	NORMAL FINDING	WARNING FINDING	RESPONSE
Digital pulse	Bilateral equal; no elevation above baseline	Asymmetric or elevated pulse	Delay training; veterinary evaluation
Limb temperature	Symmetric; no focal heat over MC3 or fetlock	Asymmetric heat or focal warmth at loading sites	Delay training; investigate source
Limb filling	Absent or minimal; bilateral symmetric	Asymmetric filling or persistent joint effusion	Extended rest; veterinary assessment
Trot-up soundness	Sound on firm surface in straight line	Any irregularity of gait at trot	Stop training; veterinary evaluation before resumption
Breeze efficiency	Speed and stride at or above prior baseline	Progressive decline across consecutive breezes	Six-race warning signal may be active; reassess

RACING 2YOs AND CAREER LONGEVITY — WHAT THE RESEARCH SHOWS

Burns (2021, PMC7916178) summarizing Jockey Club EID data: Horses that raced as 2-year-olds had more lifetime starts, more wins, greater career earnings, and more years raced than horses first racing at 3 or 4. Horses first starting at 3 or 4 years of age had greater odds of catastrophic musculoskeletal injury. The evidence does not support delaying racing to protect horses. It supports racing correctly — with architectural discipline — to build the structural foundation that makes long careers possible.

Research Foundation — Chapter 13

Race Start as ~9.3% Fatigue Life Event

Morrice-West et al. (2022). Sci Rep 12:11528. PMC9262984.

Six-Race Warning Signal

Wong et al. (2023). EVJ 55(2):194. PMC10084173. 18% MSI risk increase per 0.1 m/s speed decrease.

Racing as 2YOs and Career Longevity

Burns (2021). Animals 11(2):463. PMC7916178.

Race Spacing and Structural Risk

Crawford et al. (2021). PMC7910838. Verheyen et al. (2006). EVJ 38(4):310.

CHAPTER FOURTEEN

Transitioning Into the Three-Year-Old Season

The three-year-old season is not a fresh start. It is the first test of everything the two-year-old year built.

Two Pathways to Injury — And Why the Transition Matters

Research has identified two distinct mechanisms by which bone fatigue leads to injury in racehorses, and both are directly relevant to the two-to-three-year-old transition.

The first pathway occurs when high-speed loading is introduced too rapidly to a poorly adapted skeleton. This is the primary injury risk of the early two-year-old season. The KTM architecture is designed to avoid this pathway through phased escalation and timed reset blocks.

The second pathway occurs when training and racing continue for a period that exceeds the recovery capacity of a well-adapted skeleton. This is the primary injury risk of the later career — and the direct consequence of carrying unresolved fatigue from a compressed two-year-old season into escalating three-year-old demands. A horse that arrives at three structurally depleted has already entered the second pathway before the season begins.

THE TWO-PATHWAY FRAMEWORK

Pathway 1: Bone damage from loading introduced too rapidly in a poorly adapted skeleton. Characteristic of early career injuries. Prevented by: phased escalation, foundation period, repetition-before-escalation. Pathway 2: Bone damage from training and racing exceeding the recovery capacity of a well-adapted skeleton. Characteristic of mid-to-late career injuries. Prevented by: six-week reset blocks, race spacing discipline, Phase VI reset before the three-year-old season. The 2-to-3-year-old transition is the pivot point between pathways. Architecture determines which one the horse is on.

The Condition of Entry

A properly executed KTM two-year-old season concludes with Phase VI — a full six-week structural reset that clears accumulated fatigue, allows the remodeling cycle to complete, and positions the horse at the start of its three-year-old campaign from the same biological platform it had at the start of Phase V competition. The account is balanced. Structural margin is preserved.

Without that reset, the three-year-old year begins with latent fatigue layered into bone that is now being asked to absorb greater loading at longer distances. A horse completing four two-year-old race starts without a subsequent reset arrives at its three-year-old season having consumed roughly 37% of fatigue tolerance before its first three-year-old start.

1.37	1.79
fatal injuries per 1,000 starts for 2-year-olds	fatal injuries per 1,000 starts for 3-year-olds — a 31% increase at the transition year

Distance Progression at Three

Three-year-old racing frequently extends beyond the sprint distances dominating the two-year-old calendar. A horse moving from six-furlong two-year-old racing into mile-and-sixteenth three-year-old prep races is absorbing roughly 40% more cumulative stride cycles per race start. Research confirms that longer race distance is an independent risk factor for catastrophic musculoskeletal injury. Distance escalation at three must follow the same architectural discipline as at two: progressive, deliberate, and supported by adequate structural foundation.

2-TO-3-YEAR-OLD DISTANCE ESCALATION: STRUCTURAL IMPLICATIONS

DISTANCE STEP-UP	STRIDE CYCLE INCREASE	STRUCTURAL REQUIREMENT	KTM PREPARATION STATUS
6f → 7f	~16%	Adequate cortical tolerance at sprint; Phase VI reset completed	Met by KTM Phase V and Phase VI
6f → 1 mile	~33%	Full cortical adaptation; no latent 2YO fatigue	Met only if Phase VI reset completed
6f → 1 1/16 mi	~42%	Maximum demand at transition; 2YO fatigue amplifies risk significantly	Requires Phase VI plus progressive distance build before routing
5f → 1 mile+ (skipping distances)	~65%+	Exceeds progressive escalation principle; structural deficit guaranteed	NOT recommended — always include intermediate distance preparation

Reintroducing High-Speed Work After Phase VI

After the Phase VI reset, the return to high-speed work follows the same sequence that governed the return from Phase IV — single works building progressively through 3/8, 1/2 and 5/8. One work per week throughout. The first week back carries the highest vulnerability.

THREE-YEAR-OLD RETURN-TO-WORK PROTOCOL AFTER PHASE VI RESET

RETURN WEEK	WORK TARGET	BASIS
Week 1 (most vulnerable)	Slow galloping.	Remodeling porosity resolving. New bone matrix still maturing. Do not interrupt.
Week 2	Single easy 3/8 at :37+.	First mechanical signal back — easy controlled pace. Not a performance test.
Week 3	1/2 mile at :49+.	First work back at 1/2 — controlled.
Week 4	5/8 mile at 1:01+.	First work back at competition distance — controlled.
Week 5+	Normal 3YO preparation resumes.	Full 3YO training under KTM density accounting rules.

THE THREE-SEASON ARCHITECTURAL TRAJECTORY

SEASON	KTM LABEL	STRUCTURAL PURPOSE	WHAT IS BUILT
Two-Year-Old	**Structural Programming**	Maximize cortical adaptation during peak modeling responsiveness; deliver horse to 3YO with preserved structural margin	Cortical geometry, cross-sectional strength, fatigue tolerance baseline, aerobic platform
Three-Year-Old	**Competitive Expansion**	Extend structural platform into longer distances and higher stakes; maintain density discipline under competitive pressure	Distance tolerance, cumulative fatigue management at higher volume, career earnings foundation
Four-Year-Old	**Durability Dividend**	Harvest structural investment made at 2 and 3; compete at peak ability from a fully consolidated skeleton	Peak competitive output, injury resilience, full career earnings potential

The three-year-old season is where the two-year-old blueprint is read. A horse that arrives consolidated, structurally sound, and properly reset does not face the three-year-old season as a threat. It faces it as an opportunity. Architecture determines which one it is.

Research Foundation — Chapter 14

Two Pathways to Injury

Wong et al. (2023). Front Vet Sci 10:1260554. EVJ 55:194, PMC10084173.

Fatal Injury Rate by Age

Burns (2021). PMC7916178. 1.37 / 1.79 / 1.86 per 1,000 starts for ages 2 / 3 / 4+.

Longer Race Distance as Independent Risk Factor

Hitchens et al. (2019). Vet J 245:29. Morrice-West et al. (2022). Sci Rep 12:11528.

Return-to-Work Vulnerability

Carrier et al. (1998). JAVMA 212(10):1582.

Career Longevity and 2YO Racing

Tanner et al. (2013). EVJ 45:20. Velie et al. (2013). EVJ 45:410.

CHAPTER FIFTEEN

The Four-Year-Old Dividend

Speed builds bone. Density builds fatigue. Sequencing builds durability. The four-year-old year is where sequencing pays.

What Architecture Purchases

The KTM Model Program is not designed to produce cautious two-year-olds. It is designed to produce durable four-year-olds. Every architectural decision in the two-year-old year — every reset block, every density constraint, every repetition-before-escalation requirement — is a down payment on a four-year-old campaign from which full career earnings can be realized.

Peak Racing Age: What the Research Shows

Gramm and Marksteiner (2010) analyzed the relationship between age and racing performance across large populations of Thoroughbreds and found that peak racing ability occurs at approximately 4.45 years of age. The improvement from age two to peak represents approximately 10 lengths in sprint distances and 15 lengths at route distances. Subsequent analysis of Japanese Racing Association data across more than 400,000 starts confirmed this pattern. The four-year-old year is where Thoroughbred talent fully matures — but only if the structural foundation has been preserved to support it.

CAREER OUTCOMES — HORSES RACING AS 2YOs vs. FIRST STARTING AT 3 OR 4

OUTCOME METRIC	RACED AS 2YO	FIRST RACED AT 3 OR 4	SOURCE
Lifetime starts (ages 3+)	Significantly more starts from age 3 onward ($p<0.001$)	Fewer total career starts	Tanner et al. 2013
Career length (years raced)	Longer career on average	Shorter career	Velie et al. 2013
Career earnings	Higher lifetime earnings	Lower lifetime earnings	Chin et al. 2025
Catastrophic MSI risk	Lower odds of catastrophic injury	Greater odds of catastrophic MSI	Burns 2021
4YO peak realization	Structural platform intact to reach peak at 4.45 yrs	Structural deficit may limit 4YO performance if escalation was compressed	Gramm & Marksteiner 2010

The Bone Geometry That Cannot Be Recovered

After the modeling window closes at the end of the two-year-old year, the geometric advantage of early cortical loading cannot be meaningfully increased. The cross-sectional area, cortical thickness, and structural stiffness built during the two-year-old year are the foundation from which the entire career is built. A four-year-old competing on a well-adapted skeleton from its two-year-old year is competing with a structural advantage that was earned two years earlier — and can no longer be earned.

Consistency as a Structural Health Indicator

Wong et al. (2023) found that horses sustaining musculoskeletal injuries showed measurable decline in speed and stride length beginning approximately six races before injury. For the four-year-old operating within a well-managed career, consistency of performance across starts is not just an athletic indicator — it is a structural health indicator. A horse maintaining its speed and stride characteristics across consecutive starts at four is demonstrating that its structural margin remains intact.

CONSISTENCY AS STRUCTURAL CONFIRMATION

18% increased MSI risk per 0.1 m/s racing speed decrease across consecutive starts. 11% increased MSI risk per 10 cm stride length decrease across consecutive starts. A four-year-old horse maintaining consistent speed and stride figures across a competitive campaign is not merely performing well. It is demonstrating that the structural account funded by the two-year-old architecture remains in surplus. Declining figures are not a performance problem to be trained through. They are a structural signal to be respected.

The Architecture–Economics Connection

ARCHITECTURAL DECISIONS AND THEIR FINANCIAL CONSEQUENCES

ARCHITECTURAL DECISION	SHORT-TERM COST	LONG-TERM FINANCIAL CONSEQUENCE
Foundation period before first timed work	Delayed competitive debut	Lower injury rate in first racing season; horse available for more starts at 3 and 4
Repetition before escalation at each distance	Slower breeze progression	Structural tolerance confirmed at each level; competition distance reached safely
Phase IV six-week reset before competition	6-week delay before first race start	Horse enters competition structurally sound; lower first-year injury risk
4-week race spacing standard in Phase V	Fewer race starts per season	Each start made from partial consolidation; fatigue accumulates at manageable rate
Phase VI reset before 3YO season	Season ends 6 weeks earlier	Zero fatigue debt carried to 3YO; full structural margin available for longer distances

The Financial Case — What Injuries Actually Cost

Beyond race earnings, the KTM program makes a straightforward financial argument that every owner and farm manager can understand: an injured horse costs money from every direction simultaneously. Veterinary bills, lay-up fees, rehabilitation time, the training investment already spent getting the horse to competition fitness — all of it stops producing return the moment the horse is off the track. And when that horse eventually comes back, it must be rebuilt from scratch, burning through another training cycle before it can compete again.

The two Active Reset blocks in the KTM program are not expensive interventions. They are six weeks of daily jogging and easy galloping — the lowest-cost period in the entire training calendar. Compare that to the cost of a single significant musculoskeletal injury: lay-up fees alone can run for months, veterinary care is substantial, and the horse returns to training having lost all its competitive fitness. The reset blocks that prevent that outcome cost a fraction of what a single injury costs to manage.

The KTM program does not eliminate risk — no training program can. But it systematically reduces the probability of the fatigue-related injuries that are preventable, and fatigue-related injuries account for approximately 78% of all racing fatalities and career-ending outcomes. Protecting that 78% is where the financial case is made.

The four-year-old dividend is not a bonus. It is a return on a two-year investment. Speed builds bone. Density builds fatigue. Sequencing builds durability. Architecture is the difference between a career that peaks at four and one that ends at three.

Research Foundation — Chapter 15

Peak Racing Age — 4.45 Years

Gramm, M. & Marksteiner, R. (2010). PMC4013968. Improvement from 2YO to peak: ~10 lengths sprints, ~15 lengths routes.

Career Longevity Data

Tanner, J.C. et al. (2013). EVJ 45:20. Velie, B.D. et al. (2013). EVJ 45:410.

Catastrophic MSI Rate by Age at First Start

Burns, T.A. (2021). PMC7916178. Inverse association between age at first start and career length.

Genetic Analysis — Career Earnings and Career Length

Chin, E.W.M. et al. (2025). NZ J Agricultural Research. 26,920 NZ Thoroughbreds across 17 seasons.

Consistency as Structural Health Indicator

Wong et al. (2023). EVJ 55:194. PMC10084173. 18% MSI risk per 0.1 m/s speed decrease.

Structural Ledger

KTM conceptual framework

The cumulative account of structural deposits and withdrawals made by every training decision across the two-year-old season. Deposits include progressive high-speed work at appropriate intervals, six-week Active Resets, and repetition before escalation. Withdrawals include compressed escalation, stacked density without recovery, and race starts without adequate consolidation. When deposits consistently exceed withdrawals, structural margin grows. When withdrawals dominate, fatigue debt accumulates — and compounds into the three-year-old year.

Structural Margin

KTM operational term

The reserve of bone fatigue tolerance remaining above the threshold at which injury risk escalates. A horse with high structural margin can absorb the loads of competition without approaching its fatigue limit. A horse with depleted structural margin is operating with little or no reserve — the next hard work or race start may be the one that tips the balance.

2YO / 3YO / 4YO

Age designations used throughout

Two-year-old, three-year-old, four-year-old. Used to refer to the horse's age and the corresponding competitive season. The KTM program governs the 2YO season in full and establishes the structural foundation from which the 3YO and 4YO campaigns are built.

ANATOMICAL AND BIOLOGICAL TERMS

Third Metacarpal Bone

MC3 — the cannon bone

The primary load-bearing bone of the front limb between the knee (carpus) and the fetlock joint. The most common site of fatigue-related injury in the Thoroughbred racehorse, particularly at the condyles — the rounded articular ends that form the fetlock joint surface.

Peak compressive strain at racing speed: 4,000–5,000 microstrain. Source: Nunamaker et al. (1990) J Orthopaedic Research 8(4):604.

Subchondral Bone

The bone immediately beneath joint cartilage

Dense cortical bone that supports the cartilage surface of the fetlock and other joints. The primary site of fatigue microdamage in racehorses and the anatomical location measured in the Whitton (2014) and Morrice-West (2022) research studies.

Osteoclasts

Bone-removing cells

Specialized cells that resorb (break down) damaged or fatigued bone tissue, creating space for new bone to be deposited. Osteoclast activity is suppressed by high-speed cyclic loading — which is why the repair cycle does not activate during continuous race training.

Osteoblasts

Bone-building cells

Specialized cells that deposit new collagen matrix which subsequently mineralizes into organized, stronger bone. Osteoblasts follow osteoclasts in the remodeling sequence. Their activity is what produces the labeling coefficient reading in bone histology studies.

Osteocytes

Bone sensor cells

Mature bone cells embedded within cortical tissue that detect mechanical deformation under load and send chemical signals to recruit either osteoclasts or osteoblasts. The primary sensing mechanism by which bone responds to exercise-induced strain.

Bone Modeling

Active construction of new bone geometry

The process by which bone changes its size, shape, and cross-sectional geometry in response to new loading demands. Dominant in young horses — peaks during the two-year-old year. Requires sufficient strain magnitude to activate. Cannot be stimulated by jogging or slow work alone.

Bone Remodeling

Ongoing repair and maintenance of bone tissue

The process by which old or damaged bone tissue is replaced with fresh, organized bone. Active throughout life. Requires time and removal of high-speed loading to complete. Suppressed during continuous race training. The biological mechanism targeted by the KTM Active Reset blocks.

Microdamage / Microfracture

Microscopic cracks in cortical and subchondral bone

Normal byproduct of high-speed cyclic loading. Each racing stride produces bending, compression, and torsional forces that create microscopic cracks in bone tissue. Microdamage is the cellular trigger for repair — it is expected and normal. Injury occurs when microdamage accumulates faster than remodeling can resolve it.

Cortical Bone

The dense outer shell of bone

The hard, compact outer layer of long bones including the cannon bone. Cortical thickness and cross-sectional geometry are the primary structural variables built during the two-year-old modeling window. These geometric advantages cannot be meaningfully recovered after the modeling window closes.

Porosity Window

The mid-reset vulnerability period

The period during weeks 2–3 of an Active Reset when bone is temporarily more porous than before rest began. Osteoclasts have removed damaged tissue but osteoblasts have not yet completed the replacement. This is the most structurally vulnerable point of the reset period. Sub-threshold jogging and easy galloping are safe during this window; high-speed work is not.

Source: Martig et al. (2014) Vet J 202(3):566. PubMed 24528139.

Fatigue Life

The structural endurance budget of bone

The total number of loading cycles a bone can sustain before the probability of failure rises to an unacceptable level. Fatigue life decreases exponentially with increasing load magnitude. Each race start consumes approximately 9.3% of subchondral bone fatigue life at the primary loading site.

Source: Morrice-West et al. (2022) Sci Rep 12:11528. PMC9262984.

RESEARCH AND STATISTICAL TERMS

Musculoskeletal Injury

MSI

Any injury to the bones, joints, tendons, or ligaments of the limb caused by training or racing load. In the research cited throughout this program, MSI specifically refers to injuries serious enough to result in lost training days, veterinary intervention, or career-ending/fatal outcome.

Labeling Coefficient

LC — direct measure of active bone formation

A laboratory histomorphometry measurement quantifying how much new bone is actively being deposited at a given site. Measured in the Whitton (2014) study by examining sections of the palmar metacarpal condyle. Training horses: 0.24/mm. Resting horses: 0.85/mm. The threefold difference is the biological proof that high-speed loading suppresses the repair cycle.

Source: Whitton et al. (2014) Vet J 202(3). PubMed 25296852. P<0.001.

Odds Ratio

OR — a measure of relative risk

A statistical measure expressing how much more likely an outcome is in one group versus another. An OR of 1.0 means equal risk. An OR of 8.05 (Crawford 2021) means horses in their first 10–14 weeks of training were 8.05 times more likely to sustain an MSI than horses past that window.

Crawford et al. (2021) Animals 11(1):270. PMC7910838. OR 8.05, 95% CI 1.92–33.76, p=0.004.

Confidence Interval

CI — statistical range of certainty

The range within which researchers are statistically confident the true population value lies. A 95% CI of 1.92–33.76 means researchers are 95% confident the true odds ratio falls somewhere between those values. Wide CIs reflect smaller study populations; narrow CIs reflect larger ones. The direction of effect remains valid regardless of CI width.

Bone Mineral Content

BMC

The concentration of calcium and other minerals within bone tissue, measurable on radiographs using an aluminum penetrometer. Higher BMC generally indicates stronger, denser bone. One of the key markers used to assess whether a training program is building or depleting structural integrity over time.

Differentially Expressed Genes

DEGs — measure of tissue biological transition

Genes that are actively switching on or off in response to a biological change. The 136 DEGs identified in the 2024 JAVMA transcriptome analysis are genes associated with skeletal muscle assembly that upregulate during the two-to-three-year-old transition. Their activity during Phase VI is one of the four biological objectives of the second Active Reset.

Back Scattered Scanning Electron Microscopy

BSEM — bone histology imaging method

The laboratory imaging technique used in the Whitton (2014) study to measure bone resorption and formation activity at the microscopic level. Not a clinical tool — a research measurement technique used on post-mortem bone samples to quantify the labeling coefficient and bone area fraction.

Positron Emission Tomography

PET — nuclear imaging of bone metabolism

Advanced nuclear imaging that detects areas of increased bone metabolic activity based on radioisotope uptake. Referenced in Chapter Five: Burns (2021) reported that 65% of horses with fetlock lesions showed measurable improvement on PET imaging after a 12-week rest period.

Equine Injury Database

EID — Jockey Club national injury and career records

The Jockey Club's national database of Thoroughbred racing injuries and career performance records. The Burns (2021) career longevity data — showing that horses racing as 2YOs had more lifetime starts, higher earnings, and lower catastrophic MSI risk — was drawn from EID records covering tens of thousands of horses.

Source: Burns, T.A. (2021) Animals 11(2):463. PMC7916178.

Japan Racing Association

JRA

Japan's central horse racing authority, overseeing all major Japanese racing. Referenced in Chapter Fifteen: analysis of 400,000+ JRA starts confirmed the Gramm and Marksteiner (2010) finding that peak Thoroughbred racing ability occurs at approximately 4.45 years of age.

Source: Gramm, M. & Marksteiner, R. (2010) PMC4013968.

LOAD THRESHOLD QUICK REFERENCE

The following thresholds are the operational decision boundaries used throughout the KTM program. All are derived from population-level research and represent risk inflection points — not absolute guarantees.

KTM PROGRAM — KEY THRESHOLDS AND THEIR RESEARCH SOURCES

THRESHOLD	VALUE	MEANING	SOURCE
HSE pace	≤15 sec/f	Any work at this speed or faster counts toward the 60-dayHSE total	Morrice-West 2020
Safe HSE zone	< 20f / 60 days	Adaptation exceeds microdamage. Optimal range during foundation and escalation phases	Estberg 1996
Alert HSE zone	20-34f/60 days	Microdamage accumulating faster than remodeling. Monitor; do not escalate	Estberg 1996
Danger HSE zone	> 35f / 60 days	3.9× fatal injury risk. Remodeling suppressed. Reduce density immediately	Estberg 1996
MSI high-risk window	Wks 1–14	OR 8.05 for MSI in first 10–14 weeks of training — highest-risk period in 2YO career	Crawford 2021
Race start fatigue cost	~9.3% per start	Estimated subchondral bone fatigue life consumed per race start	Morrice-West 2022
Race spacing — preferred	28+ days	Adequate partial consolidation. Standard KTM target	KTM framework
Race spacing — monitor	21–27 days	Minimal consolidation. Requires clean structural monitoring	KTM framework
Race spacing — avoid	< 14 days	No consolidation possible in 2YO season. Structural deficit compounds	KTM framework
Six-week reset minimum	6 weeks	Minimum duration for remodeling suppression to fully resolve after high-speed loading removed	Whitton 2014
Speed decline MSI signal	0.1 m/s decrease	18% increased MSI risk per 0.1 m/s racing speed decrease across consecutive starts	Wong 2023
Stride decline MSI signal	10 cm decrease	11% increased MSI risk per 10 cm stride length decrease across consecutive starts	Wong 2023
Peak racing age	~4.45 years	Age at which Thoroughbred racing ability peaks — the target the KTM architecture is built to reach	Gramm 2010

All thresholds are operational guidelines derived from population-level research. Individual horse responses may vary. Veterinary oversight is recommended throughout.

Every number in this reference section has a research anchor. Every threshold is a biological reality, not an arbitrary rule. The architecture works because the biology demands it.

KTM MODEL PROGRAM

46-Week Two-Year-Old Training Calendar

A week-by-week visual roadmap of the complete KTM two-year-old season. Every phase, every work, every decision point — laid out so trainers, veterinarians, and owners can see the full program at a glance. Use this calendar as your operational reference throughout the season.

Phase I Foundation	Phase II Breeze Prog.	Phase III Active Reset I	Phase IV Comp Prep	Phase V Competition	Phase VI Active Reset II

RACE START | MAINT. BREEZE | CONFIRMED | POROSITY WINDOW

WK	PHASE	WEEKLY ACTIVITY AND TRAINING NOTES
1	PHASE I	Horse arrives — broken to saddle and begun ridden work. TACK WALKING ONLY. Acclimation to track environment, stable routine, surroundings.
2	PHASE I	TACK WALKING ONLY. Continue acclimation. No physical demands beyond learning the environment and daily routine.
3	PHASE I	LIGHT JOGGING begins IF trainer satisfied AND veterinary monitoring agrees. Conditional — horse and vet dictate timing. Below modeling threshold.
4	PHASE I	LIGHT JOGGING continues. Assess progression. Vet monitoring ongoing. Transition to easy galloping when horse going consistently well.
5	PHASE I	JOGGING + EASY GALLOPING. Transitioning from jog to gallop as horse settles. Aerobic base building.
6	PHASE I	EASY GALLOPING. Horse should be galloping consistently. Continue building aerobic base. Prepare for introduction of 1/8 mile open gallops.
7	PHASE I	1/8 MILE OPEN GALLOPS begin — 1st increment. Open gallop, not timed. Initiates bone remodeling. Begins prevention of bucked shins. Bridge to breeze work.
8	PHASE I	1/8 MILE OPEN GALLOPS — 2nd increment. Building effort across the week. Bone mineral content increasing. Easy galloping on off days.
9	PHASE I	1/8 MILE OPEN GALLOPS — 3rd increment. Consistent open gallop efforts. Bone responding to load. Easy galloping on off days.
10	PHASE I	1/8 MILE OPEN GALLOPS — 4th increment. Final preparation week before first breeze. Horse should be galloping freely and confidently. Vet clearance before Phase II.
11	PHASE II	BREEZE: 1/8 mile. Easy open effort — establish rhythm. Gallop-out 1f easy. Single introductory breeze. Not repeated.
12	PHASE II	BREEZE: 1/4 mile (target :26–:27). First breeze at modeling threshold pace. Gallop-out 1f easy. Establishes bone-building signal. Note any heat or filling post-work.
13	PHASE II	BREEZE: 3/8 mile (target :37+). First exposure — controlled. Repetition rule begins here. Gallop-out 1f. Must confirm before advancing.

WK	PHASE	WEEKLY ACTIVITY AND TRAINING NOTES
14	PHASE II	BREEZE: 3/8 mile CONFIRMED (target :36). Structural tolerance confirmed at 3/8. Gallop-out 1f. CONFIRMED — advance to 1/2 mile next week.
15	PHASE II	BREEZE: 1/2 mile (target :49+). First exposure — controlled. Gallop-out 1–2f. Confirm before sharpening. Monitor limbs carefully post-work.
16	PHASE II	BREEZE: 1/2 mile CONFIRMED (target :48). Structural tolerance confirmed at 1/2. Gallop-out 1–2f. CONFIRMED — advance to 5/8 mile next week.
17	PHASE II	BREEZE: 5/8 mile (target 1:01+). First exposure at competition distance — controlled. Gallop-out 1–2f. Confirm before sharpening.
18	PHASE II	BREEZE: 5/8 mile CONFIRMED (target 1:00). Structural tolerance confirmed at competition distance. Gallop-out 1–2f. CONFIRMED — step back next week.
19	PHASE II	STEP-BACK BREEZE: 1/2 mile (target :49). Deliberate density reduction before Active Reset I. Protects structural account. Gallop-out 1f. Active Reset I begins next week.
20	RESET I	ACTIVE RESET I BEGINS. Daily tack walking, jogging and easy galloping. High-speed loading removed. Osteoclast recruitment begins. Remodeling suppression starts to resolve.
21	RESET I	ACTIVE RESET I — Week 2. Daily tack walking, jogging and easy galloping. Horse in visible daily work for vet oversight. Resorption phase underway.
22	RESET I	■ POROSITY WINDOW — Week 3. Resorption phase peaks. Bone temporarily more porous than before reset. Easy galloping continues. No high-speed loading under any circumstance.
23	RESET I	■ POROSITY WINDOW — Week 4. Light galloping only. Osteoblasts beginning replacement phase. Bone repair progressing.
24	RESET I	ACTIVE RESET I — Week 5. Bone formation phase active. Easy galloping — extend distance as consolidation progresses. Fitness maintained throughout.
25	RESET I	ACTIVE RESET I — Week 6. Consolidation completing. Easy galloping continues. Mineralization consolidating. Vet assessment recommended before Phase IV begins.
26	COMP PREP	PHASE IV BEGINS — Return to work. FIRST BREEZE BACK: easy 3/8 at :37+. First mechanical signal after reset. Confirm bilateral soundness before advancing. Daily jogging and galloping continues.
27	COMP PREP	COMPETITION PREP: 1/2 mile at good pace (target :49+). Gallop-out 1–2f. Confirm before advancing to 5/8. Structural monitoring protocol active.
28	COMP PREP	COMPETITION PREP: 5/8 mile (target 1:01+). Gallop-out 1–2f. Competition distance confirmed. Horse structurally ready for race selection. Target first race in Week 29 or when right fit appears.
29	PHASE V	RACE START #1. Race + gallop-out = full HSE event (~9.3% fatigue life consumed). BEGIN RECOVERY: tack walking, jogging, easy galloping only. Structural monitoring: digital pulse, limb temp, limb filling, trot soundness.
30	PHASE V	POST-RACE RECOVERY. Tack walking, jogging and easy galloping. Daily structural monitoring continues. Assess recovery progress.
31	PHASE V	DAY 10 POST-RACE: MAINTENANCE BREEZE — 1/2 mile (target :49). Standard between-race breeze. Confirms structural soundness and maintains conditioning before next race prep.
32	PHASE V	RACE PREP: 5/8 competition prep breeze (target 1:00) if structural monitoring indicators confirm readiness. Gallop-out 1–2f. Final prep before Race #2.

WK	PHASE	WEEKLY ACTIVITY AND TRAINING NOTES
33	PHASE V	RACE START #2 (4-week spacing from Race #1). Race + gallop-out. BEGIN RECOVERY immediately. Structural monitoring protocol. Cumulative fatigue: ~18.6% fatigue life consumed.
34	PHASE V	POST-RACE RECOVERY. Tack walking, jogging and easy galloping only. Daily structural monitoring. Assess recovery — any anomaly extends consolidation period.
35	PHASE V	DAY 10 POST-RACE: MAINTENANCE BREEZE — 1/2 mile (target :49). Confirms soundness and maintains conditioning. Monitor for six-race warning signals: declining speed or stride across starts.
36	PHASE V	RACE PREP: 5/8 breeze if all monitoring indicators normal (target 1:00). Gallop-out 1–2f. Final prep before Race #3.
37	PHASE V	RACE START #3 (4-week spacing from Race #2). Race + gallop-out. BEGIN RECOVERY. Cumulative fatigue: ~27.9% fatigue life consumed. 30-day HSE total tracking essential.
38	PHASE V	POST-RACE RECOVERY. Tack walking, jogging and easy galloping only. Daily monitoring. DECISION POINT: 3 starts completed — evaluate whether a 4th start fits within density accounting rules before Active Reset II.
39	PHASE V	DAY 10 POST-RACE: MAINTENANCE BREEZE — 1/2 mile if 4th start planned. OR begin transition to Active Reset II if 3 starts completed and structural account warrants protection.
40	PHASE V	OPTIONAL RACE START #4 (if 4-week spacing achieved and structural monitoring clean). Cumulative fatigue ~37.2%. After this start Active Reset II is mandatory regardless of competitive opportunities.
41	RESET II	ACTIVE RESET II BEGINS. Daily tack walking, jogging and easy galloping. High-speed loading removed. Competition fatigue begins to resolve. Remodeling cycle reactivating.
42	RESET II	ACTIVE RESET II — Week 2. Daily tack walking, jogging and easy galloping. Horse in visible daily work for vet oversight. Resorption phase underway. Funds 3YO structural account.
43	RESET II	ACTIVE RESET II — Week 3. Easy galloping continues. Remodeling cycle completing. Skeletal muscle transition (136 DEGs) active during this period. Fitness maintained throughout.
44	RESET II	ACTIVE RESET II — Week 4. Easy galloping extends in distance. Consolidation progressing. Bone formation markers approaching baseline. 3YO structural account building.
45	RESET II	ACTIVE RESET II — Week 5. Easy galloping. Consolidation nearing completion. Structural margin approaching surplus. Begin planning 3YO season prep schedule.
46	RESET II	ACTIVE RESET II — Week 6 / FINAL WEEK. FIRST BREEZE BACK: easy 3/8 at :37+ — return-to-work signal. STRUCTURAL ACCOUNT REPLENISHED. 3YO season begins from surplus. Zero fatigue debt carried forward.

KEY DECISION POINTS — WHEN TO PAUSE AND ASSESS

End of Week 10: Veterinary clearance recommended before Phase II breeze work begins. **Weeks 13 and 14:** If horse does not confirm clean at 3/8 after two works, hold at 3/8 — do not advance. **Weeks 15 and 16:** Same rule at 1/2 mile — repetition must confirm before advancing to 5/8. **Weeks 17 and 18:** Same rule at 5/8 — both works must be clean before step-back and reset. **Week 22–23:** Porosity window — no high-speed loading under any circumstance. **Week 25:** Veterinary assessment recommended before Phase IV return to work. **Post-Race Monitoring:** Any anomaly in digital pulse, limb temperature, filling, or trot soundness extends the recovery period. Do not advance on schedule if indicators are outside normal range. **End of Phase V:** After 3–4 starts, Active Reset II is mandatory regardless of competitive opportunities or horse form.

46 weeks. Six phases. One biological sequence. The calendar is not the constraint. The biology is the constraint. The calendar is the tool we use to respect it.

www.ingramcontent.com/pod-product-compliance
Lightning Source LLC
LaVergne TN
LVHW022026200926
848481LV00007B/2564